Place names on
maps of Scotland
and Wales

First published 1968. Reprinted 1970, 1973 and 1978.
Reprinted with additions and corrections 1981.
Reprinted 1987, 1990.

ISBN 0-319-00223-3

CONTENTS

	Page
Foreword	4

Part I — Gaelic

Abbreviations	5
Pronunciation	5
The Definite Article	6
Glossary	6
Common Anglicisations	12

Part II — Scandinavian

| Glossary ... ... | 14 |

Part III — Welsh

Abbreviations	16
Pronunciation	16
Accent	16
Consonant Changes	17
Glossary	18

FOREWORD

Many years ago a small pamphlet entitled "The most common Gaelic words used on the Ordnance Survey Maps" was issued for use with one-inch maps of Scotland, and in 1935 the Board of Celtic Studies, the University of Wales, compiled a small booklet covering "The most common Welsh words used on the Ordnance Survey Maps" for users of maps of Wales.

This latter booklet was reprinted in 1949. At about the same time it was decided to increase the scope of the pamphlet on the Gaelic names; this was completely recompiled by the Royal Scottish Geographical Society and was published in a much enlarged edition in 1951.

The present edition is essentially a combination of revised versions of the two earlier publications and so the style is not completely consistent throughout. For example, some of the abbreviations used in the Gaelic and Welsh sections are different and there are differences in layout in the two pronunciation guides.

The Ordnance Survey is indebted to Dr. W. F. H. Nicolaisen of the School of Scottish Studies (University of Edinburgh) for his advice on the Gaelic and Scandinavian elements, to Mr. Wm. Matheson (Senior Lecturer in the Dept. of Celtic, University of Edinburgh) for the revised guide to Gaelic pronunciation, and to the Board of Celtic Studies (University of Wales) for help given with the production of the Welsh glossary.

PART I - GAELIC

ABBREVIATIONS

E = English	adj. = adjective	dim. = diminutive
G = Gaelic	f. = feminine gender	id. = idem, the same
L = Latin	g. = genitive case	lit. = literally
OE = Old English	m. = masculine gender	pron. = pronounced
ON = Old Norse	n. = nominative case	q.v. = quod vide, which see
Sc. = Scots	pl. = plural	s.v. = sub verbo, under the word
W = Welsh	sg. = singular	v. = vide, see

PRONUNCIATION

The vowels are broad — a (ah), o (aw, oh), u (oo); and narrow — e ea ei (ay), i (ee). They may be long (` ´) or short, with ò equal to aw, ó equal to oh, è and é as in French and eu next to a nasal equal to è, otherwise equal to é.

c next to a broad vowel equals c in cove; next to a narrow vowel equals k in king.

g next to a broad vowel equals g in gone; next to a narrow vowel equals g in give.

d next to a broad vowel equals d + th in and that (spoken rapidly); next to a narrow vowel equals d in duke.

t next to a broad vowel equals t + th in that thong (spoken rapidly); next to a narrow vowel equals t in tune.

l in all positions and ll in medial and final positions, next to a broad vowel, equal l + th in although (spoken rapidly); l in initial position and ll in medial and final positions, next to a narrow vowel, equal l in Italian miglia; l in medial and final positions, next to a narrow vowel, equals l in elect, fail.

n in initial position and nn in medial and final positions, next to a broad vowel, equal n + th in than that (spoken rapidly); n in medial and final positions next to a broad vowel, equals n in honour, man; n in inital position and nn in medial and final positions, next to a narrow vowel, equal gn in French seigneur; n in medial and final positions, next to a narrow vowel, equals n in mine, pin; except that n in final position, following ai, oi, ui, equals second nn above.

r in initial position and rr in medial and final positions, whether next to a broad or narrow vowel, equal r in ream, roam; r in medial and final positions, whether next to a broad or narrow vowel, equals r in every, hoary.

s next to a broad vowel, equals s in sod; next to a narrow vowel, equals sh in shed; initial s is silent after an t-.

ch next to a broad vowel, equals ch in Scots loch.

ch next to a narrow vowel, equals ch in Scots dreich, German ich.

gh and dh next to a broad vowel, like g in go, with tension relaxed so that the stop becomes a spirant; next to a narrow vowel, like y in yew, but usually silent in medial and final positions.

th in initial and medial positions equals h in hot, hit; practically silent in final position.

sh equals h in hot, hit.

bh and mh equals v in vole, veil.

fh is silent.

ph equals f in fan, fin.

Note:—h is inserted after initial consonants (other than d t l n r s) of masculine nouns in the genitive singular to indicate aspiration following the definite article; pronunciation as shown above.

THE DEFINITE ARTICLE IN GAELIC

SINGULAR

Masculine
Nominative an, am (before b, f, m, p), an t- (before vowels).

Genitive a', an (before d, t, l, n, r, fh), an t- (before s followed by a vowel or s followed by l, n, r).

Feminine
a', an (before d, t, l, n, r, fh, and s followed by b, g, m, p, t), an t- (before s followed by a vowel or s followed by l, n, r).

na, na h- (before vowels).

PLURAL

Masculine and Feminine
na, na h- (before vowels).

nan, nam (before b, f, m, p).

GLOSSARY

(Reference to Latin, English, Old English, Old Norse, Scottish and Welsh words indicates that the Gaelic words in question are loan-words from these languages)

abhainn,g. aibhne............................*river*
acairseid, acarsaid, ON akkarsæti...*anchorage*
achadh...*field*
achlais..*arm-pit*
agh,g. aighe,*pl.* aighean..............*hind, heifer*
aifreann, L offerenda,.................(1) *mass;*
    (2) in place-names, *a gift to the church*
ail, aileach......................*rock or stony place*
àilean.............*green spot, enclosure, meadow*
airbhe, eirbhe......*a dividing wall or boundary*
aird, g. airde,..................*height, promontory*
àirigh.....■..*shieling*
aiseag..*ferry*
aisir, aisridh...................*pass between rocks*
aitionn..*juniper*
àlainn..*beautiful*
allt..*burn, stream*
alltan,...................................*little stream*
amar,g. amair..............*trough, channel, ditch*
amhach...*neck*
annaid............................*a mother church*
aodann,g. aodainn..............................*face*
aoidh, see uidh
aoineadh..........*steep brae with rocks, moraine*
aonach......................*moor or market place*
arbhar...*corn*
ard, aird.......................*height, promontory*
àth.............................*ford(m.); kiln(f.)*

bac,g. b(h)aic,*pl.* bacaichean, ON bakki, bank, ridge.....................*bank, peat-bank*
bachall, L baculum..................*staff, crozier*
bad,g. b(h)aid,..........*tuft, clump of trees or shrubs, also a place*
bàgh,g. b(h)àigh, E bay............................*bay*
bàidhte............*drowned, liable to flooding, liable to drowning (e.g., sheep)*
baile,g. b(h)aile.....*town or hamlet, homestead*
bàn,f. bhàng. bhàin(m.), bàine(f.), *pl.*bàna...............................*fair, white*
ban-rìgh..*queen*
bard,g. b(h)aird........................(1) *poet;*
    (2) *enclosed meadow* (Sc. ward)
barpa (Hebrides and Skye)......*rude conical heap of stones, sepulchral cairn*
barr.....................................*top, summit*
bàta,g. b(h)àta,*pl.* bàtaichean, OE bàt, ON bàtr......................................*boat*
bàthaich,g. b(h)àthaich, bàthcha.......*byre, sanctuary* (in deer forests), *shelter*
beag,f. bheag,g. bhig(m.), bige(f.), *pl.*beaga.....................................*little*
bealach,g. b(h)ealaich........................*pass*
bealaidh,g. b(h)ealaidh....................*broom*
bean,g. mnà, *n.pl.*mnathan, g.*pl.*ban....................*wife, woman*
bearn...*gap*
beinn,g. beinne,g.*pl.* beann..........*mountain*
beithe..*birch*
beithir,g. b(h)eithir.....*serpent, wild beast, monster*

beul,g. b(h)eòil............................mouth
biast,g. béiste.....................beast, monster
bidean.....................................pinnacle
bile...lip, edge
binnean..............small and peaked mountain
biolair................................water-cress
biorach......(1) adj. sharp-pointed; (2) noun,
      dog-fish; heifer; young horse
blàr,g. b(h)làir.............cleared space, plain
bó,g. bà..cow
boc,g. b(h)uic..................................buck
bodach,g. b(h)odaich..........old man; spectre
bodha, ON boði, a breaker....sunk rock in sea
bog,g. bhuig, buige...(1) adj., soft, (2) noun,
            soft place
bonn.......................................base, foot
borrach............................rough hill grass
both, bothan.................primitive stone or
          turf house; bothy
bradan..salmon
bràighe,g. b(h)ràghad................upper part
breac,f. bhreac,g. bhric(m.), brice(f.);
 pl.breaca..(1) adj., speckled; (2) noun, trout
breug,g. bréige, adj.breugach..........a lie;
 false; applied to stone cairns or standing
 stones which, at a distance, may bear
 resemblance to a human form; e.g.,
 Buachaille Bréige, false shepherd; Fir
 Bhréige Chalanais, the false men of
         Callernish, Lewis
breun..foul
broc,g. b(h)ruic..............................badger
brochan,g. b(h)rochain....literally gruel or
 porridge, but applied in place-names to
 anything broken up or comminuted,
 as: Coire Bhrochain (Cairngorms), the
      corry of the broken stones
bruach,g. bruaich.....................bank, brink
bruthach,g.b(h)ruthaich.....steep place, brae
buachaille.................................herdsman
buaile......................................cattle-fold
buidhe..yellow
buidheanaich.........................yellow place
bùirich........................roaring, bellowing
bun..........root; in place-names generally
 applied to the mouth of a river or
 stream as in Bonawe, G Bun Abha,
 mouth of the river; Bunchrew (Inverness),
    G Bun Chraoibhe, near the tree

cabar...antler
cabhsair..............................E causeway
cabhail..................................fishing creel
cachaileith.................field gate or hurdle
cadha..........................steep place; pass
caibeal, L capella...........................chapel
caigeann,......................pair, couple, brace
cailleach,g. caillich, L pallium...........nun;
         old woman; hag
caiplich............................place of horses

cairdh............................weir; fish-pound
caiseal, L castellum.......stone wall, stone fort
caisteal, L castellum......................E castle
cala or caladh.............................harbour
call, coll..hazel
calltuinn...hazel
calman, calaman, L columba................dove
cam,g. chaim, caime......................crooked
camas..............................channel, bay,
    in inland places a bend
canach...............................cotton grass
car..bend
carach..winding
caochan..streamlet
caol, caolas............narrow, strait, firth, kyle
caorann............rowan tree or mountain ash
capall, L caballus..................horse or mare
carn,g. c(h)uirn,c(h)airn.....heap of stones,
    applied to round rocky hills
carr or càthair..rough or broken mossy ground
carraig, W carreg.............................rock
cas,g. chais, caise............................steep
cat..cat
cath...battle
cathair,g. cathrach......circular stone fort;
       chair; fairy knoll
ceann,g. c(h)inn..................head, headland
ceap,g. c(h)ip, L cippus....................block
ceapach.................................tillage plot
cearc,g. circe....................................hen
cearc-fhraoich...................moor hen, grouse
cearcall, L circulus.................circle, a hoop
ceard..............................craftsman, smith
ceardach..........................smithy, forge
ceathach,g. c(h)eathaich....................mist
ceathramh................................quarter
chora,g. caorach........................... sheep
cill,g. cille,g.pl. ceall, L cella..........church,
        burying place
cìoch,g. cìche................................nipple
clach,g. cloiche...............................stone
clachan....place of stones, applied primarily
 to a stone house, especially a cell or
 church; secondarily to a "Kirktown",
      hamlet; burying-place
cladach,g. c(h)ladaich...............shore, beach
cladh...................................burial-ground
claidheamh..................................sword
claigeann..........(1)m. skull, head, rounded
   hillock; (2)f. in-field of townland
clais..........furrow, narrow and shallow valley
clamhan............................kite, buzzard
clàr.....................flat stretch of ground;
 place where crossings were made on planks
cléireach, L clericus.......................cleric
cleit, ON klettr......................rock, cliff
cluain....................green plain, pasture
cnàimh,g. c(h)nàmha.......................bone
cnap, ON knappr............................hillock
cnoc,g. c(h)nuic.......................round hill

còig..five
coileach..cock
coille...................................wood, forest
còinneach......................................moss
còinneachan........................place of moss
coirce..oats
coire........a round hollow in mountain side;
           cirque; sea-gulf, whirlpool
coitcheann....................common pasture
comar.....................................confluence
comraich..................................sanctuary
con, choin v. cù
corr,g. curra..................................crane
corran........sickle; low cape tapering to a point
còs..nook
crann..tree
craobh,g. craoibhe..........................tree
crasg,g. c(h)raisg, c(h)roisg..........a crossing
Creachann.........the bare wind-swept place
                   about the top of a hill
creag,g. creige..............crag, rock, or cliff
creamh...................................wild garlic
crìoch,g. crìche......................... boundary
critheann.........................the aspen tree
crò,g. c(h)ròtha, pl. cròithean..sheep cot, pen
cròc.................................deer's antler
cròcach......................branched, branching
crodh,g. c(h)ruidh..........................cattle
croich,g. croiche, L cruc-em...........gallows
crois...E cross
croit..E croft
crom,g. chruim, cruime................crooked
crosg,v. crasg
cruach..........................heap, stack, bold hill
cù,g. c(h)oin,g.pl. con......................dog
cuach..........................cup-shaped hollow
cuan..sea
cuidhe, ON kví..........................fold, pen
cuil. ....nook, recess
cuilc..reed
cuileann,g. c(h)uilinn.....................holly
cuithe, L puteus..........pit, wreath of snow,
                                  narrow glen
cùl...............................back, hill-back
cumhann,g. chumhainn, cuinge.....narrow,
                                            strait

dà..two
dabhach, dobhach.......(1) large vat or tub;
   (2) (in place-names) large measure of land
dail,g. dalach....................field, dale, haugh
damh,g. daimh............................ox, stag
darach..oak
dearcag..berry
dearg,g. dheirge, deirge....................red
deas..south
dìg..moat
dìollaid, diallaid...........................saddle

dìridh........................climbing, ascending
dìseart, L desertum............hermit's retreat
dìthreabh....................desert, wilderness
dobhar..water
dobhran...otter
doire..grove
dòirlinn.................................isthmus,
              usually covered at high water
domhain..deep
donn..brown
dorcha..dark
dorn...fist
dris..bramble
drochaid......................................bridge
droigheann, draigheann..............blackthorn
druid..thrush
druim..........................the back, a ridge
dubh,f. dhubh; g.m.dhuibh; g.f.duibhe;
   pl.dubha.....................................black
duine,pl. daoine.............................man
dùn,g. dùin........fortress, castle, heap, mound

each,g. eich.................................horse
eadar..between
eag..notch
eaglais, L ecclesia..........................church
eala,pl. ealachan..............................swan
ear..east
earb..roe
earrann............share, portion, section of land
eas,g. easa, pl.easan.................waterfall ;
              rough ravine (Perthshire)
easg(a), easgaidh, easgainn....a marsh, swamp
eidheann...ivy
eileach..................................mill-lade;
   narrow shallow stream joining two lochs;
   arrangement for catching fish in a stream
eileag..deer-trap
eilean,g. eilein,pl. eileanan,
   ON ey-land................................island
eileirg...................deer-trap, pass where
              deer were killed or captured
eilid..hind
eòrna..barley
eun,g.sg. and n.pl.eòin......................bird

fada..long
fadhail, ON vaðill............ford in sea channel
faire.......................................watching
faithir, foithir..................shelving declivity
fàl,g. fàil.........................hedge, palisade
fàn.......................gentle slope, level space
fang,g. f(h)aing, fainge, Sc. fank..sheep-pen,
                                           fank
faoileag....................................sea gull
faoilinn..beach
fas, fasadh............stance, station, level spot
fasgadh.......................................shelter
feadag..plover

feadan,g. f(h)eadain...........*narrow glen or hollow; streamlet*
feannag.....................(1) *crow;* (2) *lazy-bed*
fear,g. f(h)ir................................*a man*
fearann..................................*land, estate*
fearn, fearna............................*alder tree*
féith,g. féithe...*bog, slowly moving stream, bog channel* (lit. *vein, sinew*)
feur,g. f(h)eòir..............................*grass*
feusag...*beard*
fiacail...*tooth*
fiadh,g. f(h)éidh..............................*deer*
fiodhag...................................*bird-cherry*
fionn.....................*white, fair; blessed, holy*
fireach..*hill ground*
fitheach..*raven*
fliuch...*wet*
fraoch..*heather*
frìth, OE frið............................*deer-forest*
fuar...*cold*
fuaran,g. f(h)uarain.............*well, spring; secondarily, a green spot*

gabhar,g. g(h)abhair, gaibhre, *also*
  gobhar,g. g(h)obhair, goibhre..........*goat*
gaineamh,g. gainimh, gaineamha(i)ch....*sand*
gall,g. g(h)aill, g(h)oill.....*stranger, lowlander*
gamhainn,g. g(h)aimhne, g(h)amhna......*stirk*
gaoth..........................(1) *wind;* (2) *marsh*
garaidh..*den*
garbh,f. gharbh; g.m.ghairbh, g.f.gairbhe;
  pl.garbha....................................*rough*
garradh,g. g(h)arraidh, ON garðr...*wall or dike; garden*
gart.. *enclosure*
gead.......................*rig, narrow strip of land*
gèadh,g. g(h)eòidh..........................*goose*
geal,f. gheal; g.m.ghil,g.f.gile;
  pl.geala......................................*white*
gearr....................(1) *adj. short;* (2) *noun*, *also* gearrsaich, *a hare*
gearraidh, ON gerði....*strip of land between moor and plain, outer pastures* (Lewis)
geata...E *gate*
geodha, geò, ON gjà..................*chasm, rift*
gil, ON gil....................*ravine; water-course*
gille... ...*boy, lad*
giubhas, giuthas..................................*fir*
giuran.......(1) *cow parsnip;* (2) *barnacle goose*
glac,g. glaice.................................*hollow*
glais...*stream*
glas,f. ghlas; g.m.ghlais, g.f. glaise;
  pl.glasa.............................*grey, green*
gleann,g. g(h)linne.............*narrow valley, dale, glen*
gob,g. g(h)uib...........................*point, beak*
gobha, gobhainn....................*blacksmith*
gobhal..*fork*
gobhar,v. gabhar
gobhlach,g. g(h)obhlaich..................*forked*

gorm,f. ghorm; g.m.ghuirm,
  g.f.guirme; pl.gorma..............*green, blue*
gort, gart, dim. goirtean......*field, enclosed orn-land*
grannda...*ugly*
grian,g. gréine..................................*sun*
grianan....................................*sunny hillock*
guala, gualainn..................*shoulder of a hill*

iar..*west*
imrich..........................*removing, flitting*
inbhir.......*place of meeting of rivers, where a river falls into the sea or lake, confluence*
inghean, inghinn...........*daughter;* v.*nighean*
inis,g. in(n)se......(1) *island;* (2) *meadow by the side of a river; haugh;* (3) *resting place for cattle, etc.*
iobairt.....*offering, sacrifice* (of offerings or gifts of land made to the church)
iochdar...................................*lower part*
iodhlann (iolann),g. iodhlainn...*corn-yard, granary*
iola.....................*fishing rock, fishing bank*
iolair,g. iolaire...............................*eagle*
iomaire.........................*rig, ridge of land*
ionga...*nail*
iosal,g. ìsle.......................................*low*
iubhar,g. iubhair..............................*yew*

labhar..*loud*
lacha.....................................*wild duck*
lag,g. luig...................................*a hollow*
lagan,g. lagain........................*little hollow*
làir,g. làire, làrach..............................*mare*
làirig...*a pass*
lann,g. lainne.. ....*enclosure, land*
laogh,g. laoigh.......................................*calf*
làrach..........................*site of ruined house*
leaba, leabaidh,g. leapa..................*bed, lair*
leac,g. lice..........................*flat stone, slab*
leacach.....................................*stony slope*
leamh(an)..*elm*
lèana,dim. lèanag, lianag.....*wet plain or lea*
learg,g. leirge........................*plain, hillside*
leathad,g. leathaid, leothaid........*slope, brae*
leathan(n),g. leathain(n)..................*broad*
leitir,g. leitire, leitreach....*slope, side of a hill*
leth......*half;* e.g., Leth-allt, *half-burn,* i.e., *burn with one steep side*
leum,g. leuma..................................*leap*
liath,g. léith, léithe....................*grey, blue*
linn, linne..*pool*
lìon,g. lìn..*flax*
lios, leas........................*enclosure, garden*
loch,g. locha; dim.lochan,g. lochain...*lake; lakelet*
loinn, locative case of lann,q.v.
lòn,g. lòin...............*marsh, morass, pool, meadow* (in Skye *a slow stream*)
long,g. luinge..*ship*

losg, loisgte..........................*burnt ground*
losgann,g. losgainn..........................*frog*
luachair,g. luachrach......................*rushes*
lùb,g. lùib...................................*a bend*

machair,g. m(h)achaire, mach(a)rach....*plain*
madadh,g. m(h)adaidh............*dog, wolf, fox*
magh,g. moighe, muighe............*plain, field*
maigheach,g. maighiche......................*hare*
màm,g. m(h)àim........*large, round or gently
   rising hill; mountain gap or pass*
manach,g. m(h)anaich, L monachus......*monk*
mang...*fawn*
maol,g. maoile............................*bare top*
marc..*horse*
marg...*merk-land*
màs..*buttock*
meadhon,g. m(h)eadhoin.................*middle*
meall,g. m(h)ill..............................*lump,
   applied to a round hill*
meanbh..............................*small, slender*
meann,g. m(h)inn................................*kid*
mèinn,g. mèinne..................................*ore*
meur,g. m(h)eòir, meòire...*finger; branch,
   applied to small streams*
mìn,g.f. mìne.................................*smooth*
mnà,g.s., mnathan,n.pl., of bean,q.v.
mòd,g. m(h)òid, ON mót.........*court, trial,
   meeting*
mòine,g. mòna, mònach....................*peat*
mòinteach.....................*mossy ground, moor*
mol, mal, mul, ON möl,g. malar..*shingly beach*
molach................................*rough, shaggy*
monadh, W mynydd.......*hill, mountain, moor*
mór,f. mhór;   g.m. mhóir;   g.f. móire;
   pl.móra..............................*large, great*
muc,g. muice......................................*pig*
muileann,g. m(h)uilinn, muilne,
   L molinum......................................*mill*
muir,g. m(h)ara............................*the sea*
mullach,g. m(h)ullaich..............*top, summit*
mult, molt,g. m(h)uilt....................*wether*
muran,g. m(h)urain......................*sea bent*

nathair,g. nathrach,pl.
   nathraichean....................*serpent, adder*
nead,g. nid..*nest*
neimhidh.................................*church land*
nighean,g. nighinne.................*daughter,
   young woman; v.*inghean

òb,g. òba, òib, ON hóp........................*bay*
odhar,g. odhair, uidhir, uidhre..*dun-coloured*
oidhche...*night*
oitir,g. oitire............................*sand bank*
òr,g. òir..*gold*
ord,g. uird..........................(I) *hammer;*
   (2) *in place-names, round hill*
òs,g. òsa, òis, ON ós...*river mouth or outlet;
   in Lewis, slowly moving water*

paire,g. pairce, Sc. park.....................*field*
pait,g. paite,pl. paitean..........*hump, ford,
   stepping-stones*
peighinn, ON penningr................*pennyland*
peit, pit, Pict. petia................*croft, stead,
   share, portion*
ploc,g. p(h)luic..............*lumpish promontory*
poll,g. p(h)uill....................*pool or pit; mud*
pòr,g. p(h)ùir...............................*pasture*
port,g. p(h)uirt.............*port; harbour, ferry*
portan, partan..................................*crab*
preas,g. p(h)ris............*bush, clump, thicket*

raineach,g. rainich...........................*fern*
ràmh,g. ràimh......................................*oar*
rann, roinn, pl. ranna....................*divison*
raon..*field*
ràth.......................................*circular fort*
rathad, g. rathaid, rothaid............*road, way*
reamhar...................................*thick, fat*
rèidh.........................*smooth, level, plain*
reilig, roilig, L reliquiae...............*cemetery*
riabhach.........................*brindled, greyish*
riasg,g. réisg...*morass with sedge or dirk-grass*
rìgh...*king*
rinn, roinn, pl reanna.......*point, promontory*
roid...*bog myrtle*
ròn,g. ròin..*seal*
ros,g. rois......................*promontory; wood*
ruadh...................................*red, brown*
rubha..........................*promontory, cape*
ruighe,g. ruighe(adh)..........(I) *fore-arm;*
   (2) *in place-names, slope, run for cattle,
   shieling*
sabhal, sobhal,g. sabhail, sobhail..........*barn*
sac,g. saic................................*horse load*
sagart,g. sagairt, L sacerdos...............*priest*
saidh,g. saidhe................................*bitch*
sàil,g. sàile......................................*heel*
sailean.....................*arm of the sea, inlet*
sàl, sàil,g. sàile...................*salt water, sea*
salach......................................*foul, dirty*
samh...*sorrel*
saobhaidh.................................*fox-den*
saor,g. saoir...................*carpenter, wright*
seabhag, OE heafoc..........................*hawk*
seahdc...*seven*
seagal..*rye*
sealg,g. seilge.....................*hunt, hunting*
seamrag,g. seamraige......*trefoil or shamrock*
sean..*old*
seangan,g. seangain........................*ant*
seanrach,g. searraich................*foal, colt*
seileach,g. seilich..........................*willow*
seipeil.......................................*E chapel*
sgadan,g. sgadain...........................*herring*
sgarbh,g. sgairbh, ON skarfr.........*cormorant*
sgeir,g. sgeire, ON sker.........*sea rook, skerry*
sgìre, sgìreachd, OE scir...................*parish*
sgitheach,g. sgithich...................*hawthorn*

sgoilt(e)...*split*
sgorr, sgurr...............................*rocky peak*
sian. slon,g. slne..............................*storm*
sìdh, sìth...................*fairy knoll, hillock;*
  dim. sìdhean,g. sìdhein
sionnach,g. sionnaich..........................*fox*
sleac,g. slic.... *flat stones* (Badenoch; v. leac)
sleamhainn...............................*slippery*
sliabh,g. sléibhe........*mountain, face of a hill*
slios,g. sliosa...........................*side, slope*
sloc,g. sluic............................*pit, hollow*
snàmh,g. snàimh.....................*swimming*
sneachd(a)....................................*snow*
soc,g. suic..*snout*; whence Socach, Socaich,
  projecting place, mossy ground between
  fork of streams, often anglicised as
  Succoth
spardan...*roost*
speireag,g. speireig............*E sparrow-hawk*
spidean...................................*v.* bidean
spréidh,g. spréidh, *L* praeda............*cattle*
sràid, *L* strata *(via)*..................*road, street*
srath.......................................*strath, valley*
sròn,g. sròine..........................*nose, point*
sruth,g. srutha........................*current,*
  stream; dim. sruthan,
  streamlet
stac,g. staca, *ON* stakkr........*steep conical hill*
stob...*point*
stùc..*peak*
subh,*pl.* suibhean.......................*raspberry*
suidhe.....*seat, sitting or resting place; level*
  shelf in a hillside
sùil,g. sùla...*eye*

taillear, g. tailleir..............................*tailor*
tairbeart...........*a crossing, portage, Isthmus*
talamh,g. talaimh, talmhainn................*land*
tana..................................*thin, shallow*
tarbh,g. tairbh...............................*bull*
tarsuinn.........................*transverse, across*
teampull,g. teampuill, *L* templum....... *temple*
  or church

teanga........ ....*tongue*
tearmann, tearmad,
  L termon-is....................*,sanctuary, girth*
teine....................................*fire, beacon*
tigh,g. tighe...................................*house*
tiobar,g. tiobairt.........................*well*
tioram..*dry*
tìr...*land*
tòb..............*a bay* (Lewis); same as òb,q.v.
tobar,g. tobair..................................*well*
tobhta, *ON* topt........*ruin with walls standing*
todhar,g. todhair.....*manure, dung bleaching*
toll,g. tuill.....................*hole;* as *adj. hollow*
tolm,g. tuilm, *ON* hólmr......*an island in a*
  river or near the shore
tom,g. tuim.......*knoll or rounded eminence*
  (in the east); *copse* (in the west)
tòn,g. tòin........................*buttock, haunch*
torc,g. tuirc............................*boar*
torr,g. torra.....................*heap. hill, castle*
tràigh,g. tràighe, tràghad.........*(tidal) beach*
trì...*three*
train...*third*
tuath..*north*
tulach,g. tulaich.................*knoll, hillock*
  (there were also forms tiolach. tealach,
  often the source of anglicised forms)
tunnag,g. tunnaig...........................*duck*

uachdar,g. uachdair.............*top, upper part*
uaine..*green*
uamh,g. uamha..............................*cave*
uchd..............*breast, ascent; dim.*uchdan,
  short steep bank
uidh, *ON* eið....................*isthm*us*, ford.*
  slowly moving water
ùig, *ON* vík, bay, creek...*a nook, hollow or bay*
uinnseann...................................*ash tree*
uisge...*water*
ulaidh...*..treasure*
ùruisg...................*monster, goblin,brownie*

COMMON ANGLICISATIONS OF GAELIC NAMES
(The anglicised forms are those listed first)

Anglicised	Gaelic
ach	achadh
ald, alt	allt
annat, annet	annaid
ari, -ary	àirigh
arn	earann
arra, arrie, errie	airbhe, eirbhe
atin	aitionn
auch	achadh
auchter	uachdar
auld, ault	allt
bal, balla, bally	baile
balloch	bealach
bane, vane	bàn, bhàn
bar	barr
bea, bae	beithe
beg	beag
ben	beinn
blair	blàr
bon	bonn *and* bun
bowie	buidhe
brae, bread	bràighe, bràghaid
breck	breac
brock	broc
buie, vuie	buidhe, bhuidhe
cairn	carn
cambus	camas
carrick	carraig
carry	caraidh
caw	cadha
chipper	tiobar
clack	clach
claddach	cladach
corrie, corry	coire
coul	cùl
cow	call, coll
craig	creag
crask	crasg
dal, dall, dalloch, del,	dail, g. dalach
darroch	darach
derg	dearg
der(ry)	doire
doch	dabhach, dobhach
dour	dobhar
dow, du	dubh
drain	draigheann
drem, drim. drom, drum	druim
dysart	dìseart
eder	eadar
edin	aodann
egles, eglis	eaglais
elan, ellan	eilean
eldrick, eldrig, elrick	eileirg
faoghail	fadhail
fassie	fas, fasadh
ferin	fearann
fern	fearna
fin	fionn
foss	fasadh
-four	pòr, g. p(h)ùir
garve	gar, garbh
glen	gleann
glack	glac
goul	gobhal
gour, gower	gobhar
gow, gown, gowan	gobhann
ibert	Iobairt
inch, insh	inis
inver	inbhir
ken, kin	ceann
keppoch	ceapach
kil	usu. represents cill; but may stand for caol, cùl, cùil and coille
knap	cnap
knock	cnoc
kyle, kyles	caol, caolas
lagg, laggan	lag, lagan
leck	leac
letter	leitir
lia	liath
lin, lynn	linn
lis	lios
logie	lag(aidh)
luib	lùb
maddy	madadh
more	mór, sometimes Moire
moy	magh
muck	muc
ochter	uachdar
pen, penny	peighinn
plock	ploc
pol, pul	poll
ra	ràth
reoch, rioch	riabhach
rhe, rhi, rie, ry	ruighe
ross	ros
schie	sìdh
sgor	sgorr
shan	sean
shee	sìdh
sheen	sian, sìon, g. sìne
slochd	sloc
soul	sabhal

stack	stac
strath	srath
strone	sròn
tarbert, tarbet, tarbat	tairbeart
tay, tea, ty	tigh
termit	tearmad,*s.v.* tearmann

tiber, tibber, tipper	tiobar
tilloch, tilly	tiolach,*s.v.* tulach
tober	tobar
tor	torr
tullich, tulloch, tully	tulach
ur	new, fresh, **recent**

PART II - SCANDINAVIAN
GLOSSARY
(The elements are listed in their ON forms)

á "*river*" is common in names of rivers, e.g., Laxa (Shetland); occurs also in Thurso.

bekkr "*stream*" appears as -beck.
birki "*birch forest*" occurs in Birkisco (Skye), Birksneip (Midlothian), etc.
ból "*farm*" occurs in Eribol and in Skibo (Sutherland).
bólstaðr "*dwelling-place, homestead*" occurs frequently in Shetland as -bister, -busta, in Lewis as -bost; also in Scrabster (Caithness).
breiðr "*broad*" appears commonly as Bray-; also in Breibiste (Shetland).
bryggja "*jetty, landing-place*" appears commonly as -brigg.
býr "*farm, estate*" appears as -by, -bie.

eið "*isthmus*" occurs in Scapa, Hoxa and Eday (Orkney).
ey "*island*" appears as -ey, -ay, and -a, as in Foula, Staffa, Fladda, Flotta, Hoy (ON Há-ey "*lofty isle*"), Isle of May (ON Má-ey "*gull isle*").

fjall "*rough hill*" occurs in Goatfell (Arran), etc.
fors "*waterfall*" occurs in Forse (Caithness).
fugl "*bird*" occurs in Foula.
fær "*sheep*" occurs in Fair Isle and Fara (Orkney).

garðr "*enclosure*" occurs commonly as -garth; in Shetland as -gaard.
gil "*ravine, water-course*" occurs in Gillabol (Sutherland).
gjá "*chasm, rift*" appears as geo, gio; in Gjoster (Shetland).
gnípa "*steep hill*" occurs in Neep (Shetland), Kneep (Lewis).
gnúpr "*peak*" occurs in Noup Head (Orkney) and in the Noop of Noss (Shetland).

hár "*high*" occurs in Habost (Lewis) and Hoy (Orkney).
haugr "*mound, barrow*" occurs commonly as -howe; also in Hoxa (Orkney) and Hjoag (Shetland).
hestr "*horse, stallion*" occurs in Hestaval (Lewis).
hof "*pagan temple*" occurs as Hov- in Orkney and Shetland.
holmr "*small island*" appears as Holm, Tolm.
hóp "*bay*" occurs in Longhope (Orkney).
hris "*brushwood*" occurs in Risay (Lewis).
hross "*horse*" appears as Ros(s)-. as in Rossal (Shetland),

kirkja "*church*" appears commonly as Kir(k)-, -kirk.
klettr "*rock*" appears as Clett.
konungr "*king*" occurs in Conisby (Islay), Konister, formerly a royal domain and Conningsburgh (Shetland).
krókr "*bend, crook*" appears commonly as Crook-.
kví "*fold, enclosure for cattle*" is common in Orkney and Shetland as -quoy; in its older form kwei, hwei it appears frequently in Shetland also.

langr "*long*" as in Langay.
lax "*salmon*" appears in Laxay (Islay) and Laxa (Shetland).
leir "*clay, loam, mud*" appears as Lar-, Lear-, Ler-; as in Lerwick (Shetland).
lundr "*grove*" occurs as Lund-, -lund, and in Lunnasting (Shetland).
lyng "*heather*" occurs in Lingey (Harris).

meir "*sandbank*" occurs in Melbost (Lewis), Melness (Sutherland).
mjor "*narrow*" occurs in Meavig (Harris).
múli "*promontory*" occurs commonly as Mool, Mull in Shetland.
myrkr "*dark, murky*" occurs in Murkle (Caithness).
mýrr "*swampy moorland, mire*" appeart commonly as -mire.

nes "*headland, promontory, cape*" appears commonly as Ness, -ness.

ós *"river-mouth, inlet"* appears commonly as -os, -ois.

röst *"whirlpool, strong sea current"* occurs in Sumburgh Roost.

sauðr *"sheep"* occurs in Soray (Fiannan), Soa (Coll, Tiree, St. Kilda), Soay (Skye, Harris), and Soy (Coll).

setr *"house, dwelling-place"* appears commonly as -set(t); in Shetland as -ster, as in Bruster (ON Brúarsetr *"house by the bridge"*), Gjaster (ON Gjásetr *"house by the chasm"*).

skáli *"hut, shieling"* occurs in Lanskaill (Orkney), Scalloway (Shetland).

sker *"skerry"* occurs commonly as Sker-, Skeir, Skerries.

skógr *"wood"* occurs in Birkisco (Skye).

slakki *"shallow valley, depression between two hills"* appears commonly as -slack.

slétta *"plain, level field"* appears as -sleat, -slat.

stakkr *"hill, precipitous rock"* appears as -stack, -staca.

staðr *"steading farm"* occwrs in Grimersta, Scarasta (Lewis).

steinn *"stone"* appears as Sten-, Stain-, as in Stenness (Orkney).

stórr *"big"* occurs in Papa Stour and Storholm (Shetland).

þing *"assembly"* occurs in the compound þing-völlr *"plain of assembly"* in Thingwall (Shetland), Tingwall (Orkney), Dingwall (Ross), Tinwald (Dumfries).

tunga, *"tongue, spit of land"* occurs in Tongue (Sutherland).

vágr *"bay, creek, voe"* appears as -wall, -way, vagh, as in Kirkwall, Osmondwall (Orkney), Scalloway (Shetland) and Stornoway

vík *"bay, creek"* occurs as -wick. In Gaelic areas -bhig, -vig, -aig. Cf. ùig

vrá *"nook, corner of land"* appears as Wra-, Wray-, -ray.

PART III - WELSH

ABBREVIATIONS

a	= adjective	m	= masculine	np	= noun plural
ad	= adverb	n	= noun masculine	pl	= plural
f	= feminine	nf	= noun feminine	pr	= preposition

PRONUNCIATION

c = k; it is never pronounced as s.

g = the English hard g; it is never pronounced as j.

ff and ph have the same sound, that of the English f.

th has the sound of the English th in thin.

dd has the sound of the English th in this.

ch is sounded like the German ch in nach, and like the Scotch ch in loch.

f = v.

ng has the sound of the English ng in hanging. In a few words, such as bangor, ng stands for ng + g, as in finger.

ll is a voiceless l, and bears the same relation to l as th in thin to th in this.

r is trilled like the strong Scotch r.

s = s in loss. It is never sounded z.

w and i are often consonants, sounded like the English w and y respectively, as in with and yes

a is a flat a not heard in English, but approximates the a in German nach.

e when short, has the sound of the English e in pen, and when long, has approximately the sound of a in face in Northern pronunciation.

i when long, is sounded like the Northern English i in machine. When short, as in dim, it has the same sound, and not the wider i as in English dim.

o when long or medium, as in tôn, tonau, has a sound midway between the close o of English note and the open o of English not. When short as in ton, tonnau, it is more open, tending towards the o of not.

u has the sound of the first y in English mystery.

y has two sounds. (1) In monosyllables and in the ultima it has the same sound as u. (2) In other positions it has the sound of the English y in myrtle.

w when long, equals the English oo in moon, and when short, equals the English oo in book

The vowel is short in monosyllables before p, t, c, m, ng, or before two or more consonants; as top 'top', cut 'shed, sty', lloc 'fold', cwm 'valley', llong 'ship', nant 'brook, ravine'. The vowel is long in monosyllables if it is final, or followed by b, d, g, f, dd, ff, th, ch, s; as lle 'place', mab 'son', gwlad 'region', teg 'fair', tref 'homestead, town', rhodd 'gift', rhaff 'rope', cath 'cat', coch 'red', tas 'stack, rick'. If the vowel be followed by l, n or r it may be long or short, and the long vowel is marked thus (^); e.g., dôl 'meadow', lôn 'lane', tŵr 'tower'. But i and u are not circumflexed, as they are almost always long before these consonants; and the long vowel is not marked in the common words, dyn 'man' and hen 'old'.

ACCENT

Welsh words are as a rule accented on the last syllable but one.

CONSONANT CHANGES

In Welsh, initial consonants undergo certain regular changes. There are nine mutable consonants and three mutations: the soft, the nasal and the spirant.

	Radical	p	t	c	b	d	g	m	ll	rh
Mutated	Soft	b	d	g	f	dd		f	l	r
	Nasal	mh	nh	ngh	m	n	ng	No change		
	Spirant	ph	th	ch	No change			No change		

The following points may be noted:—

(i) The initial consonant of the second element of a compound undergoes the soft mutation. Example: Brithdir = brith ('speckled, coarse') + tir ('land').

(ii) The initial consonant of the genitive is often softened after a feminine singular noun, such as llan ('church, monastery'), tref ('town, hamlet, homestead'), pont ('bridge'), rhyd ('ford'), caer ('fort'), ynys ('island, holm, river-meadow'). Examples: Llanbadarn = llan + Padarn (proper name). Tregaron = tre(f) + Caron (proper name).
The initial of a personal name in the genitive is sometimes softened in place-names even after a masculine singular noun. Examples: Tyddewi = tŷ ('house') + Dewi (proper name). Cae Ddafydd = cae ('field') + Dafydd (proper name).

(iii) The initial consonant of a feminine singular noun is softened after the article. Examples: Pen-y-dref = pen ('end, head, top') + y (definite article) + tref ('town, homestead'); Pant-y-frân = pant ('hollow, valley') + y (definite article) + brân ('crow'); Rhobell for Yr obell = yr (definite article) + gobell ('ridge between two summits').
Occasionally the initial consonant of a plural form used as a dual is softened after the article. Example: Yr Eifl = yr (definite article) + geifl (plural of gafl 'fork').

(iv) The initial consonant of an adjective is softened after a feminine singular noun. Example: Rhyd Goch = rhyd ('ford') + coch ('red').

(v) The initial consonant of a following noun is softened after the prepositions, ar ('on, over'), tan (dan) ('under'), tros (dros) ('over'). Examples: Pontardawe = pont ('bridge') + ar ('on, over') + Tawe (name of a river); Pontarddulais = pont ('bridge') + ar ('on, over') + Dulais (name of a river).

(vi) The initial consonant of a following noun undergoes the nasal mutation after the preposition yn. Examples: Llanfihangel yn Nhywyn, Llanfair yn Neubwll.

GLOSSARY

abad,n..abbot
aber,n.& nf........estuary, confluence, stream
ac(e)r,nf..acre
aderyn,pl. adar,n................................bird
adwy,nf.....................................gap, pass
acl, nf.....................................brow, edge
aelwyd,nf......................................hearth
aethnen,nf..........................aspen, poplar
afallen,nf..................................apple-tree
afon,nf...river
agen,nf................................cleft, cranny
allt, pl. elltydd,nf............hillside, cliff, wood
anner, annair,nf..............................heifer
ar, pr..................................on, upon, over
ardd, nf...................................hill, height
argae,pl.argaeau,n......mill-dam, embankment
arglwydd, n..lord
arglwyddes,nf....................................lady
argoed,n,..............................wood, grove
arian, n..silver
asgwrn,pl. esgyrn,n...........................bone
athro,n......................................teacher
aur, n..gold
awel, pl. awelon, nf.........................breeze

bach,a............................small, little, lesser
bach,pl. bachau,nf....................nook, corner
baedd,n..boar
bala,n.................................outlet of a lake
ban,pl. bannau,nf.....................peak, beacon
banc,pl. bencydd,n......................bank, slope
bangor,nf..................monastery originally
                        constructed of wattle rods
banhadlen,pl. banadl, banal,nf............broom
banhadlog,nf.........................broom patch
banw,n...................................young pig
bar,n..................................top, summit
barcud, n..............................kite, buzzard
bardd,pl.beirdd,n......................bard, poet
bargod,n........................boundary (eaves)
batin, bating (see betin, etc.)
bedd,pl. beddau,n.............................grave
bedwen,pl. bedw,nf..........................birch
bedwlwyn,n...............................birch grove
beili,pl. beiliau,n..........bailey, court before
                                a house, bailiff
beidr, nf...................................lane, path
beisgawn, nf..................stack of corn or hay
belau, bele, n..................................marten
ben,nf..................................waggon, cart
bendigaid,a...................................blessed
berw,n...............................rush of water
betin, beting, bietin, bieting,n......land the
    surface of which has been pared and burnt
betws, n...........................chapel of ease
betws, n............birch grove, slope or bank
                         covered with brushwood

beudy,n.......................cowhouse, cow-shed
bid,n........................quickset hedge
blaen,pl. blaenau,n..........end, edge; source
                        of river stream;highland
blaidd,n..wolf
bod, nf................................abode, dwelling
bol bola,n.................belly, swelling, hollow
bôn,n.................................stock, stump
boncen, boncyn, (see poncyn, poncen)
bord,nf..table
bracty,n..................................malt-house
braenar, branar, brynar,n...........fallow land
braich,n. & nf...................ridge, spur, arm
brân, pl. brain, nf..............................crow
bras, pl. breision, a........ big, fat, rich, coarse
bre, nf..hill
brenin,n.,..king
brest, nf...................................hill-breast
brith,f. braith,a..................speckled; coarse
bro,nf......................region;vale,lowland
bron, bronnydd,nf.........hill-breast(breast)
brwynen,pl.brwyn,nf..........................rush
brwynog,nf...............place of rushes, marsh
brych,f.brech,a...........................speckled
bryn,pl. bryniau,n..............................hill
bryncyn,n................................hillock
buarth,pl. buarthau,n..........farm-yard;pen
buches,nf.....................milking-fold (herd)
budr,a................................dirty, mucky
buddai,nf......................................churn
bugail,pl. bugelydd. bugeiliaid,......shepherd
bustach,pl. bustych,n....................bullock
buwch,pl. buchod,nf........................cow
bwbach,n..............................bogey,goblin
bwch,n..buck
bwla,n..bull
bwlch,pl.bylchau,n...................gap,pass
bwrdd,n..table
bwth,bwthyn,n..................cottage, booth
bychan,f. bechan,pl. bychain,a.......little,tiny
byr, ber,a......................................short

caban,n................................cottage, cabin
cad,nf..battle
cadair(cader),nf................seat, stronghold
cadfa,nf.cadlan,nf.cadle,n............battlefield
cadlas,nf.................close, court of a house
cadlys,nf.................stronghold, earthworks
cadno..fox
cadw,a..........preserved(coed cadw 'spinney')
cae,pl. caeau,n.....................field, enclosure
caer,pl. caerau,nf.................stronghold,fort
cafn,n.....................................ferry-boat
calch,n..............................lime, chalk

cam,*a*..................................*crooked, bent*
camfa,*nf*..*stile*
canol,*n* & *a*.....................................*middle*
cantref,*n*............*hundred (territorial division)*
capel,*n*...*chapel*
car,*pl*. ceir,*n*...............................*car, sledge*
carn,*pl*. carnau,*nf*.......*heap of stones, tumulus*
carnedd,*pl*. carneddau,
  carneddi,*nf*............*heap of stones, tumulus*
carrai,*pl*. careiau,*nf*............*narrow strip of*
  land (thong)
carreg,*pl*. cerrig,*nf*......................*stone, rock*
carrog,*nf*...*brook*
carw,*n*..*stag*
cas (*in* Casnewydd, etc.),*n*.................*castle*
caseg,*pl*. cesig,*nf*..............................*mare*
castell,*pl*. cestyll,*n*................*castle; small*
  stronghold; fortified residence; imposing
  natural position
cath,*nf*..*cat*
cau,*a*...............................*hollow; enclosed*
cawr,*pl*. ceiri, cewri,*n*........................*giant*
cawres,*nf*......................................*giantess*
caws,*n*..*cheese*
cawsai,*nf*....................................*causeway*
cefn,*pl*. cefnydd,*n*.............................*ridge*
cefnen,*nf*..*ridge*
cefnffordd,*nf*................................*ridgeway*
ceffyl,*pl*. ceffylau,*n*...........................*horse*
cegid,*np*..*hemlock*
cegidog,*nf*..........*place where hemlock grows*
cegin,*nf*...*kitchen*
cei,*n*..*quay*
ceiliog,*n*..*cock*
ceirch, cyrch,*n*..................................*oats*
ceirchiog,*nf*.................................*oat land*
ceisiad,*pl*. ceisiaid,*n*..........................*bailiff*
celynnen,*pl*. celyn,*nf*....................*holly tree*
celynnog, clynnog,*nf*..................*holly grove*
celli,*nf*..*grove*
cemais,*n*. from *np*.............. *bend in river,*
  or coastline
cennin,*np*..*leeks*
cerddinen, cerdinen, *pl*. cerddin,
  cerdin,*nf*..........................*mountain-ash*
cerwyn,*nf*..............................*vat,cauldron*
cesail,*nf*.........................*hollow (arm-pit)*
cest,*nf*................................*belly, swelling*
ceulan,*nf*...................................*river bank*
ceunant,*n*...............................*ravine, gorge*
ci,*pl*. cŵn,*n*..*dog*
cigfran,*pl*. cigfrain,*nf*.......................*raven*
cil,*pl*. ciliau,*n*............*retreat, recess, corner*
cilfach,*nf*..*nook*
cist,*nf*..*chest*
cistfaen,*n*..................................*stone-chest*
clais,*n*..................................*ditch, trench*
clawdd,*pl*. cloddiau,*n*.............*ditch, hedge*
cledr,*nf*...................................*stave, stake*

clegyr,*n*......................................*rock, cliff*
clip,*n*..................*precipice, crag, steep slope*
cloch,*nf*...*bell*
clochydd,*n*...................*sexton, parish clerk*
clog,*nf*................................*crag, precipice*
clogwyn,*n*.................*precipice, steep rock*
  hanging on one side
clos,*n*..........................*yard before a house*
clun,*n*..........................*river-meadow, holm*
clwt, clwtyn,*pl*. clytiau,*n*.....................*patch*
clwyd,*pl*. clwydydd,*nf*...............*hurdle, gate*
clyd,*a*...*sheltered*
cnap, cnepyn,*n*............*top, summit of hill,*
  short sharp ascent
cnol,*nf*................................*hillock, knoll*
cnwc,*n*..*hillock*
cnwch,*n*..*hillock*
coblyn,*n*.................................*goblin, imp*
coc, cocyn,*pl*. cociau,*n*............*tump, hillock*
coch,*a*..*red*
coeden,*pl*. coed,*nf*............................*tree*
coetan,*nf*...*quoit*
coetgae,*n*...........*woodfield, field recovered*
  from forest; palisaded enclosure
congl,*nf*...*corner*
colomen,*pl*. colomennod,*nf*.................*dove*
colwyn,*n*..*whelp*
collen,*pl*. cyll, coll,*nf*........................*hazel*
comin,*pl*. comins,*n*........................*common*
copa,*nf*.................................*crest, summit*
côr,*n*. & *nf*...............................*stall, crib*
cored,*n* & *nf*.......................................*weir*
corlan,*nf*................................*sheep-fold*
corn,*n*......................................*horn, peak*
cornant,*nf*......................................*brook*
cornel,*n*.& *nf*..................................*corner*
cors,*pl*. corsydd,*nf*..............................*bog*
corsen,*pl*. cyrs,*nf*..............................*reed*
cotel,*n*..*paddock*
cowrt,*n*..*yard*
craf,*n*...*garlic*
craig,*pl*. creigiau,*nf*...........................*rock*
crau (*as in* creuddyn),*n*................ *pigsty,*
  stronghold
criafolen,*pl*. criafol,*nf*............*mountain-ash*
crib,*pl*. cribau,*nf*....................*crest, summit*
crin,*a*...*withered*
crochan,*n*....................................*cauldron*
croes,*nf* & *a*.....................................*cross*
croesffordd, croesheol,
  croeslon,*n*..........................*cross-roads*
crofft,*pl*. crofftau,*nf*........................*croft*
crug,*pl*. crugiau,*n*....................*heap, tump*
crwm,*a*................................*crooked, bent*
crwth,*n*...............................*tump, hillock*
crwys,*nf* & *np*...............................*crosses*
crydd,*n*.....................*shoemaker, cobbler*
crythor,*n*..*fiddler*
cul,*a*...*narrow*
cut,*pl*. cutiau,*n*............................*shed, sty*

cwar,*pl.* cwarrau,*n*.................... quarry, rock
cwarter,*n*..quarter
cwm,*pl.* cymau, cymoedd,*n*..........valley, dale
cwmin, cymin,*pl.* cwmins, *n* ....common
cwmwd,*n*..........commote (territorial division)
cŵn (see ci)
cwrn,*pl.* cyrnau, cyrn,*n*......stack, peak, point
cwrt,*n*......................................court, yard
cwrw,*n*..ale, beer
cwt,*pl.* cytiau,*n*..............................hut, sty
cwter,*nf*..............................gutter, channel
cyd (*in tir cyd*),*a*..................common (land)
cyfair, cyfer,*n*....................................acre
cyff,*nf*..................................stock, stump
cyffin,*nf*......................boundary, frontier
cyfoeth,*n*..................territory, land (riches)
cyfrwy,*n*...ridge between two summits (saddle)
cyfyng,*a*............................narrow, strait
cymer,*pl.* cymerau,*n*....................confluence
cynefin,*n*.........................haunt, sheep-walk
cytir,*n*...common
cythraul,*n*...devil
chwaen,*nf*.........................piece of land,
                            breezy, windy place
chwarter,*n*..................................quarter
chwil, chwilen,*pl.* chwilod,*nf*............beetle
chwiler,*nf*...........................maggot, viper
chwileiriog,*nf*...land abounding in maggots
                                   or vipers
chwilog,*nf*.............land infested with beetles

daear, daer,*nf*..................................land
dafad,*pl.* defaid,*nf*...........................sheep
(da)fadfa,*nf*...............................sheep-fold
dafaty, (de)feity,*n*......................sheep-fold
dalfa,*nf*..pound
dan,*pr*..................................under, below
dâr,*pl.* deri,*nf*....................................oak
das (see tas)
deilen,*pl.* dail,*nf*................................leaf
deintur,*n*.....................tenter, tenter-hooks
delw,*nf*..image
deon,*n*...dean
derlwyn,*n*................................oak-grove
derwen,*pl.* derw,*nf*............................oak
diawl,*n*..devil
dibyn,*n*.....................steep slope, precipice
diffwys,*n*...........................precipice, abyss
din,*n*..................................hill-fortress
dinas,*n.* & *nf*.....................hill-fortress (city)
diserth,*n*...............................hermitage
disgwylfa,*nf*..............place of observation,
                               look-out point
dôl,*pl.* dolau, dolydd,*nf*..................meadow
dorglwyd, *nf*..................door-hurdle, gate
draenen,*pl.* drain,*nf*..........................thorn
draenllwyn,*n*............................thorn-bush
draw,*ad*......................................yonder
drum (see trum)
drws,*n*...................gap, narrow pass (door)

dryll,*n*.............................piece of land
drysïen,*pl.* drysi,*nf*.....................bramble
dryslwyn,*n*..........................bramble bush
du,*a*....................................black, dark
dwfn,*f.* dofn,*a*................................deep
dwfr, dŵr,*n*....................................water
dyffryn,*n*......................................valley
dylluan (see tylluan)
-dyn (*as in* treuddyn, creuddyn).....enclosure

ebol,*pl.* ebolion,*n*.............................colt
efail (see gefail)
efwr,*n*............................cow-parsnip, yew
eglwys,*nf*...................................church
eiddew,*n*..ivy
eira,*n*..snow
(ei)singrug,*n*.........heap of bran or corn husks
eisteddfa,*nf*....................seat, resting place
eisteddfod,*nf*...........meeting-place, assembly
eithin,*np*.......................................furze
eithinog,*nf*..............................furze patch
elor,*nf*...bier
ellyll,*n*...............................elf, goblin
eos,*nf*.....................................nightingale
erw,*pl.* erwau,*nf*.............................acre
eryr,*n*..eagle
esgair,*nf*...........................long ridge (leg)
esgob,*n*.......................................bishop
ewig,*nf*..hind

ffald,*pl.* ffaldau,*nf*........ sheep-fold, pound, pen
ffarm (see fferm)
ffatri,*nf*.......................................factory
ffawydden,*pl.* ffawydd,*nf*....... beech or fir tree
ffawyddog,*nf*.................... beech or fir grove
fferm,*nf*..farm
ffin,*nf*.....................................boundary
ffolt,*nf*..........................sheep-fold, pound
fforch,*nf*..........bifurcation of streams, fork
                          shape of fields, etc.
ffordd,*nf*................................way, road
fforest,*nf*...........................forest, park
ffos,*nf*...........................ditch, trench
ffridd, ffrith,*pl.* ffriddoedd,*nf*..........wood;
            frith, mountain enclosure, sheep walk
ffrwd,*nf*............................stream, torrent
ffwrnais,*nf*..................................furnace
ffynnon,*pl.* ffynhonnau,*nf*..........spring, well
fry,*ad*..above
fyr, fer, *n*...............................fir, pine

gafl,*pl.* geifl,*nf*................................fork
gafr,*pl.* geifr,*nf*...............................goat
gallt (see allt)
gardd,*pl.* gerddi, garddau,*nf*.........garden;
    enclosure or fold into which calves were
                       turned for first time
garth,*n*........................ promontory, hill;
                                    enclosure

garw,*a*...................................coarse, rough
gefail,*nf*..smithy
(g)eirw,*np*...........................rush of waters
genau,*n*...................pass, opening of a valley
giât,*nf*...gate
glan,*nf*..............................river-bank, hillock
glas,*a*...green
glas, glais (*as in* dulas, dulais),*n*. & *nf*.....brook
glo,*n*.....................................charcoal, coal
glyn,*n*..............................deep valley, glen
gobell,*nf*............ridge between two summits (saddle)
godre,*n*....................foot of hill (skirt, edge)
gof,*n*..smith
gofer,*n*..................stream, overflow of a well
gogof,*pl*. gogofau,*nf*............................cave
golau,*a*.......................................light, bright
gores(t), *n*, & *nf*...........................waste land
gorffwysfa,*nf*..........................resting place
goror, *pl*. gororion, *n*. & *nf*............ boundary
gorsedd,*nf*..............tumulus, barrow, hillock
graean,*n*..gravel
graeanog,*nf*.............................gravel bank
gris,*pl*. grisiau,*n*.................................step
gro,*n*. & *nf*.....................gravel; gravel bank
grofft,*pl*. grofftau,*nf*...........................croft
grug,*n*...................................heath, heather
grwn,*n*............ridge between two furrows, ridge in a field
gwaelop,*n*......................foot of hill (bottom)
gwair,*n*..hay
gwâl,*nf*...lair
gwal,*pl*. gwaliau,*nf*.............................wall
gwar,*nf*.........ridge, place above (nape of neck)
gwartheg,*np*.............................cows, cattle
gwas,*n*....................................youth, servant
gwastad,*a*..flat
gwastad, *pl*. gwastaden, *n*...................plain
gwastadedd,*n*....................................plain
gwaun,*pl*. gweunydd,*nf*.....moor, mountain meadow, moor-land field
gweirglodd,*nf*................hay-field, meadow
gwely,*n*...........bed, resting-place, family land
gwenith,*n*..wheat
gwerfa,*nf*..........cool spot on mountain side where cattle or sheep seek shade
gwernen,*pl*. gwern,*nf*..................alder tree
gwern,*nf*. gwernog,*nf*............place where alders grow, swamp
gwersyll,*n*................................encampment
gwialen,*pl*. gwiail,*nf*.....................withe, rod
gwiber,*nf*...viper
gwig,*pl*. gwigau,*nf*....................grove, wood
gwinllan,*nf*.....................grove (vineyard)
gwlad,*nf*..region
gwlyb,*f*. gwleb,*a*...................................wet
gwndwn (see gwyndwn)
gwrach,*nf*..............................hag, witch
gwrych,*n*...................hedge, quickset hedge
gwryd,*n*..fathom

gwter (see cwter)
gwydden,*pl*. gwŷdd,*nf*........................tree
Gwyddel,*pl*. Gwyddyl, Gwyddelod,*n*.........................Irishman
gwyddfa,*nf*.........................mound, tumulus
gwyddgrug,*nf*.................... mound, tumulus
gwylfa,*nf*............................look-out point
gwyn,*f*. gwen,*a*...............................white
gwyndwn, gwndwn,*n*.............grassland, lea
gwynt,*n*..wind
gwyrdd,*f*. gwerdd,*a*........................green
gyrfa,*nf*....................................racecourse

hafn,*nf*..................................gorge, ravine
hafod,*nf*........................summer dwelling
hafodol,*n*.........................summer dwelling
hafoty,*n*...........................summer dwelling
haidd,*n*..barley
halog,*a*...............................dirty, muddy
haul,*n*...the sun
helfa,*nf*........................hunting-ground
heli,*n*...brine
helygen,*pl*. helyg,*nf*........................willow
hen,*a*...old
hendref,*nf*.........winter dwelling, old home, permanent abode
heol, hewl,*nf*..........................street, road
hesgen,*pl*. hesg,*nf*.............................rush
hesgyn,*n*..bog
hir,*a*...long
hwch,*nf*..................................sow, swine
hwnt,*ad*......................................yonder
hwrdd,*pl*. hyrddod,*n*.........................ram
hydd,*pl*. hyddod,*n*.............................stag
hynt,*nf*..way
hysb,*f*. hesb,*a*.......................dry, dried up

iet,*nf*..gate
is,*pr*.....................................below, under
isaf,*a*..................................lower, lowest
isel,*a*...low
iwrch,*pl*. iyrchod,*n*......................roebuck

lawnd, lawnt,*nf*..................open space in woodland, glade
lefel,*nf*. & *a*....................................level
llaeth,*n*..milk
llaethdy,*n*.......................milkhouse, dairy
llain,*pl*. lleiniau,*nf*...narrow strip of land, slang
llam,*n*..leap
llan,*nf*............church, monastery (enclosure)
llannerch,*nf*......................clearing, glade
llawr,*n*.................level area of valley (floor)
lle,*n*.....................................place, position
llech,*pl*. llechau,*nf*...............slab, stone, rock
llechwedd,*nf*................................hillside
lleian,*nf*..nun
lleidr,*pl*. lladron,*n*............................thief
llety,*n*....................small abode, quarters
llethr,*nf*...slope

21

llidiard, llidiart,*pl*. llidiardau,
llidiartau,*n*...................................*gate*
llo,*pl*. lloi,*n*...*calf*
lloc,*n*..*fold*
llofft,*nf*...........................*eminence (loft)*
lluest,*n*....................*shieling, cottage, hut*
llumon,*n*..........................*stack (chimney)*
llus,*np*.....................................*bilberries*
llwch,*n*..*dust*
llwch,*pl*. llychau,*n*..............................*lake*
llwm,*f*. llom,*a*.......................*bare, exposed*
llwybr,*n*...*path*
llwyd,*a*.................................*grey, brown*
llwydrew, *n*..........................*hoar-frost*
llwyfen,*pl*. llwyf,*nf*.............................*elm*
llwyn,*pl*. llwyni, llwynau,*n*..........*grove, bush*
llwynog,*n*..*fox*
llydan,*a*..*broad*
llygad,*n*...........*source of river or stream (eye)*
llygoden,*pl*. llygod,*nf*......................*mouse*
llyn,*n*. & *nf*......................................*lake*
llys,*n*. & *nf*...............................*court, hall*
lôn,*nf*......................................*lane, road*

ma-,-fa,*nf*..............................*plain, place*
mab,*pl*. meibion,*n*.............................*son*
maen,*pl*. meini, main,*n*.....................*stone*
maenol, maenor,*nf*.....*stone-built residence
of chieftain of district, rich low-lying
land surrounding same, vale*
maerdref,*nf*...............*hamlet attached to
chieftain's court, lord's demesne (maer,
steward + tref, hamlet)*
maerdy,*n*..................*steward's house, dairy*
maes,*pl*. meysydd,*n*............*open field, plain*
magwyr,*nf*............*a wall of stones without
mortar, wall*
maharen,*pl*. meheryn,*n*..............*ram, wether*
main,*a*..................................*thin, narrow*
main (see maen)
march,*pl*. meirch,*n*................*horse, stallion*
marchog,*n*.....................*knight, horseman*
marian,*n*.......*holm, gravel, gravelly ground,
rock debris*
marl,*n*...........................*marl, chalky clay*
mawn,*n*..*peat*
mawnog,*nf*..................................*peat-bog*
mawr,*a*.......................*great, big, greater*
meidr,*n*........................*lane, path*
meillionen,*pl*. meillion,*nf*..................*clover*
melin,*nf*..*mill*
melindref,*nf*..............................*mill-town*
melyn,*f*. melen,*a*............................*yellow*
merch,*n*....................................*daughter*
merddwr,*n*.......................*stagnant water*
merllyn,*n*............................*stagnant pool*
merthyr,*n*....................*burial place, church*
meudwy,*n*....................................*hermit*
miaren,*pl*. mieri,*nf*.......................*bramble*

mign, mignen,*pl*. mignedd,*nf*....*bog, quagmire*
milgi,*n*...................................*greyhound*
miliast,*nf*...........................*greyhound bitch*
mochyn,*pl*. moch,*n*............................*pig*
moel,*nf*......................................*bare hill*
moel,*a*....................................*bare, bald*
morfa,*n*....................................*marsh, fen*
morgrugyn,*pl*. morgrug,*n*....................*ant*
moryd,*nf*....................................*estuary*
mur,*pl*. muriau,*n*..............................*wall*
murddun,*n*......................................*ruin*
mwdwl,*n*.......................*stack, cock of hay*
mwyalch, mwyalchen,*nf*...............*blackbird*
mwyn,*n*..................................*ore, mine*
mw(y)nglawdd,*n*.............................*mine*
mynach,*pl*. mynych,
mynaich,*n*......................................*monk*
mynachdy,*n*....................*monastic grange*
mynachlog,*nf*.............................*monastery*
mynwent,*nf*............................*churchyard*
mynydd,*n*....................*mountain, moorland*

nant,*pl*. nentydd, naint nannau,*n* & *nf*......
dingle, glen, ravine, brook
neuadd,*nf*..*hall*
newydd,*a*...*new*
nyth,*n*. & *nf*..........*nest, inaccessible position*

ochr,*nf*...................................*side, slope*
odyn,*nf*...*kiln*
oen,*pl*. ŵyn,*n*...............................*lamb*
oer,*a*...................................*cold, exposed*
offeiriad,*n*......................................*priest*
offeren,*nf*.......................................*mass*
ogof (see gogof)
onllwyn,*n*...............................*ash grove*
onnen,*pl*. onn, ynn,*nf*...................*ash-tree*

pabell, pebyll,*nf*...............................*tent*
padell,*pl*. pedyll,*nf*............................*pan*
pair,*n*.......................................*cauldron*
palmant,*n*.................................*paved way*
pandy,*n*..............................*fulling-mill*
pant,*n*................................*hollow, valley*
panwaun,*pl*. panweunydd,*nf*......*bog where
cotton-grass grows*
parc,*pl*. parciau, parcau,*n*.........*park, field,
enclosure*
pàs,*n*..*pass*
pell,*a*...*far*
pellaf,*a*......................................*farthest*
pen,*pl*. pennau,*n*............*head, top; end, edge*
penrhyn,*n*.............................*promontory*
pentref,*n*.........*homestead, appendix to the
real 'tref', village*
pêr,*a*..*sweet*
perfedd,*n*............................*middle, midst*
perllan,*nf*...................................*orchard*

person,n.....................................parson
perth,pl. perthi,nf...........bush, brake, hedge
pi,pl. piod,nf................................magpie
pia,n..magpie
pica,a...................................pointed, sharp
pig,nf..point
pigyn,n..point
pil,n..........pill, tidal creek on coast, sea ditch
pistyll,n............................spout, waterfall
plas,n............gentleman's seat, hall, mansion
plwm,n...lead
plwyf,n...parish
poeth,a....................................burnt (hot)
pompren (pont bren),nf........plank bridge,
                                                foot-bridge
ponc,pl. ponciau,nf.................hillock, tump
poncyn,n. poncen,nf................small hillock
pont,nf...bridge
porchell, parchell,pl. perchyll,n.....young pig
porth,n.................................gate, gateway
porth,nf..............................ferry, harbour
pren,n..tree
prys(g),n..........................copse, brushwood
pwll,pl. pyllau,n............................pit, pool

rhaeadr,nf....................................waterfall
rhandir,n......allotment, fixed measure of land
rhedynen,pl. rhedyn,nf........................fern
rhedynog,nf..............................fern-brake
rhestr,nf....................................row, series
rhingyll,n..............................herald, beadle
(r)hiniog,n...............narrow pass (threshold)
rhiw, nf. & n................................hill slope
rhos,pl. rhosydd,nf...........moor, promontory
rhyd,nf..ford

saer,pl. seiri,n..............................craftsman
saeth,pl. saethau,nf...........................arrow
Sais,pl. Saeson,n......................Englishman
sant, san, sain, pl. saint,n............saint; monk
sarff,nf...serpent
sarn,pl. sarnau,nf........................causeway
siamb(e)r, siamb(a)r,nf...................chamber
sied,n..................escheat (land), common
                                            mountain land
sigl, siglen,nf.............................quagmire
simnai, simdde,nf........................chimney
siop,nf..shop
stent,n. & nf...........................extent (land)
sticil, sticill,nf......................................stile
sugn,n..quagmire
swch,nf...ploughshare, triangular piece of land
swllt,n............................treasure (shilling)
swydd,nf.....................seat, lordship, office
sych,a..dry

tafarn,pl. tafarnau,n. & nf.................tavern
taflod, tawlod,nf............................hay loft

tafol,np...................................dock plants
tafolog,nf.................place where dock grows
tail,n...................................dung, manure
tâl,n..................................end (forehead)
talar,nf................strip at right angles to
                                        furrows, headland
talwrn,pl. talyrni,n.............bare exposed
  hill-side, open space, threshing floor, cockpit
tan, dan pr..........................under, beneath
tarren,pl. tarenni,nf.................rocky tump
tarw,pl. teirw,n...............................bull
tas,nf.....................................stack, rick
teg,a...fair
telyn,nf...harp
tew,a.......................................thick, fat
tir,n................................land, territory
tlawd,pl. tlodion,a & n............. poor, pauper
tom, tomen,nf.............................mound
ton,pl. tonnau,nf...............................wave
ton, tonnen,pl. tonnau,n. & nf..grassland, lea
top,n...top
traean,n.................................third part
traeth,n................................strand, shore
trallwng, trallwm,n.............wet bottom land
traws,a. & n.................cross, transverse;
                                        direction, district
tref,nf..................homestead, hamlet, town
tro,n.....................................turn, bend
troed,n. & nf.................base, bottom (foot)
tros,pr..over
trum,pl. trumiau,n........................ridge
trwyn,n.........................point, cape (nose)
twlc,n..hovel, sty
twlch,pl. tylchau,n....................tump, knoll
twll,pl. tyllau,n...........................hole, pit
twmpath,n..........................tump, hillock
tŵr,n...tower
twrch,pl. tyrchod,n............................boar
twyn,pl. twyni,n....................hillock, knoll
tŷ,pl. tai,n......................................houst
tyddyn, ty'n,n..............small farm, holding
  (as in ty'n-y-coed, etc.)
tyle,n. & nf..............................hill, ascent
tylluan,nf..owl
tyno,n..................hollow, plain, meadow
tywarchen,pl. tywarch, tyweirch,
  tywyrch,nf..............................sod, turf
tywyll,a............................dark, gloomy
tywyn, (towyn),n.............sea-shore, strand

uchaf,a...........................higher, highest
uchel,a..high
uwch,pr...............................above, over
uwchlaw,pr...................................above

wtra,nf..lane

y,yr,'r, (definite article)........................the
ych,pl. ychen,n.....................................ox

ŷd,*n*..*corn*
ydlan,*nf*....................................*rick-yard*
(y)menyn,*n*..................................*butter*
yn,*pr*..*in*
ynys,*pl.* ynysoedd,*nf*............*island; holm, river-meadow*
ysbyty,*n*..........................*hospital, hospice*
ysgafell,*nf*...............................*ledge, brow*
ysgallen,*pl.* ysgall,*nf*........................*thistle*
ysgallog, *a & nf*...........*full of thistles; place where thistles abound*
ysgawen, *pl.* ysgaw, ysgo, *nf*..........*elder-tree*
ysgeifiog,*nf*...................*place of elder-trees*
ysgol,*pl.* ysgolion,*nf*.........................*school*
ysgol,*pl.* ysgolion,*nf*..... *ladder, ladder-like formation on mountainside*
ysgubor,*pl.* ysguboriau,*nf*...................*barn*
ysguthan,*nf* ....*wood-pigeon*
(y)sgwd,*n* ....*waterfall*
ysgwydd,*nf*..........*shoulder, projecting part of mountain like human shoulder*
ystafell,*nf*..................*chamber, hiding-place*
ystlys,*nf*...................................*side, flank*
ystrad,*n*.............*valley, holm, river-meadow*
(y)stryd,*nf*..............................*way, street*
ystum,*nf. & n*.... ....*bend, shape*
ystwffwl,*n*..*staple*
ywen, *pl.* yw,*nf*............................*yew-tree*